–THE BRITISH HERRING INDUSTRY

The Steam Drifter Years 1900–1960

CHRISTOPHER UNSWORTH

AMBERLEY

First published 2013

Amberley Publishing
The Hill, Stroud
Gloucestershire, GL5 4EP

www.amberley-books.com

British Library Cataloguing in Publication Data.
A catalogue record for this book is available from the British Library.

ISBN 978 1 4456 1081 8

Typeset in 10pt on 12pt Sabon.
Typesetting and Origination by Amberley Publishing.
Printed in the UK.

CONTENTS

Tables

Introduction:
The Herring Industry in 1900

Herring have been caught around the coasts of the United Kingdom for centuries, but until the early nineteenth century the size of the British industry and the methods it used had changed little from medieval times. The ports of East Anglia and the Firth of Forth had long enjoyed a thriving export trade in herrings but the volume they produced was small compared to that of the Dutch herring industry. By 1800 the Industrial Revolution was transforming many British industries and new technologies and production methods were bringing increased production, but the herring industry continued to operate much as it had for centuries. Change, however was on its way.

Nineteenth Century Expansion

During the nineteenth century, the hard work and entrepreneurial skills of the Scottish herring curers brought about a huge growth in the herring industry in Scotland. The curers packed freshly landed herring in salt into barrels – 'salted herring'. By keeping strictly to the quality standards set by the Fishery Board for Scotland and by using the 'Crown Brand' trade mark, the Scots developed a consistent product that was as good as the salted herring produced by the Dutch. Previously, the herring-hungry nations of northern Europe had bought very little British salted herring because of its poor quality; the Dutch product had always been preferred even though it cost three times as much. After 1860, the Scots curers began to take part in the East Anglian herring fishery and this gave a huge boost to the English herring industry.

Several technological developments contributed towards the growth of the herring fisheries, the first of which was the invention in the 1840s of machinery that could cheaply produce cotton fishing nets. These nets were larger in area but lighter in weight than the hemp nets which they replaced. By the 1880s it was estimated that by changing to cotton nets, some fishing boats were catching five times as many herring as they had with the old nets.

In the aftermath of a severe gale in 1848, which had caused the loss of many small open fishing boats, the Washington Report had recommended that fishing vessels should be of sturdier build and should provide fishermen with shelter from the elements. In larger, safer craft, fishermen were able to sail greater distances away from land, exploit new herring stocks and carry larger catches of fish back to harbour.

Some boat owners installed a small boiler on sailing drifters in order to power the capstan. This eased the heavy work of hauling in the line of nets and was another factor contributing to larger catch sizes – with a steam capstan you could work with a longer train of nets than could be hauled in by hand.

The Natural History of the Herring

Herring are an amazing species and understanding their life cycle is essential to catching them. They are a pelagic fish species, meaning that they live in the area between the sea bed and its surface, as opposed to demersal species such as sole, cod and haddock, which live permanently at or near the seabed. For much of the year herring live in large groups, 'shoals' or 'schools', in the deeper parts of the Continental Shelf. Once a year, shoals of adult herring (aged 3 years and above) migrate considerable distances to shallower waters to spawn (breed). Some herring groups spawn in the summer, some during autumn and some in the winter. Each of these groups has its own feeding and spawning areas and its own annual migration route. In the weeks before they spawn, herring feed up on the plankton that gathers just under the surface of the sea; it is the urge to feed that brings them into contact with the drift-nets which are set near the surface. In their pre-spawning state, herrings are at their highest nutritional value – the discerning Continental customers preferred their herring to be in this prime condition. After spawning, the 'spent' herring rapidly lose condition and are a less saleable product.

Herring shoals containing up to four billion fish have been recorded and a shoal of this size could be up to two miles in length. The shoals could generally be relied upon to re-appear in the same area each year, although occasionally they did not turn up or, if they did, the numbers were much smaller than usual. From very early times the three factors that made the herring an attractive quarry for fishermen were, firstly, their predictable arrival each year; secondly, their large numbers; and thirdly, they often came close to land.

Herring are an important part of the marine food chain. Natural predators include seals, dolphins, porpoises, whales, cod, hake, dogfish, seabirds and conger eels. Research in Norway has shown that an adult killer whale can eat 400 herring a day.

The Branches of the Herring Trade

Herring is a fish which deteriorates rapidly if it is not preserved by salting, smoking or freezing. By 1900, most of the British herring catch was being sold overseas as barrels of salted herring (also known as 'pickled', 'cured' or 'white' herring). This trade accounted for around eighty per cent of all the herring landed in the United Kingdom. Herring preserved by this method were still edible six months later. By the 1900s the British consumed very little salted herring, except in some Scottish fishing villages where it was a staple food in winter. Salted herring is an acquired taste – one less than enthusiastic British writer described it as like 'chewing a salty dishcloth, full of fish bones'.

Herring can also be preserved by smoking. Merchants in Yarmouth and Lowestoft had for centuries exported to Mediterranean countries hard-smoked herrings such as 'Red Herring'. These herrings might spend two to three weeks in a smokehouse, exposed to the smoke from slow-burning oak sawdust. British consumers preferred a lighter smoked herring such as the bloater or the kipper, which were smoked for just one day. A further product of this branch of the industry was the manufacture of savoury spreads such as bloater paste, once a favourite filling for sandwiches.

The arrival of the railways brought improved distribution networks and the demand for fresh herrings grew, particularly in the expanding industrial cities. In Lowestoft, the railway developer Samuel Morton Peto famously promised local fishing interests that if they supported his railway plans, the fish they landed in the morning would be on sale in Manchester the following day – and they were.

At the end of the 1890s, a new herring export trade was established with the German ports of Bremen and Altona (near Hamburg). This was the 'Klondyke' trade – so named (with a slight misspelling) after the Alaskan gold rush of 1897. It involved packing whole (ungutted) herring into large wooden crates together with a mixture of ice and salt. These were sent by fast steamer across the North Sea and when they were unloaded 24 hours later in Germany, they could still be sold as fresh herring or they could be further processed, cured or marinated. The Klondyke trade took large quantities of British herring (it accounted for 10 per cent of total herring exports) and, by providing competition to the Scots curers in the quayside markets, it helped ensure that the fishermen obtained reasonable prices for their herring.

From the early 1900s some manufacturers successfully produced herrings in tin cans, generally cooked in tomato sauce, and names such as 'Tyne Brand' of North Shields and 'Jenny' of Lowestoft would soon became household favourites.

A Geographic Advantage

At different times of the year, herring were found in commercial numbers in the North Sea, the Moray Firth, the Minches, the Firth of Clyde, the Irish Sea and the

English Channel. Because the herring grounds were close to land, British fishermen were generally able to land their catches of herrings within hours of catching them. Their French and rivals, who had to fish further away from their home ports, used larger drifters with a crew of up to thirty men (compared to the British drifter with a crew of eight to ten) which remained at sea for several weeks. As they hauled and emptied their nets, they commenced the salting and barrelling process on board the vessel. On arrival at their home port, the herrings were repacked and then marketed. The fact that the British herring were caught, gutted, properly salted and packed in barrels within 24 hours was the reason that Russian and German buyers preferred them.

Geography also accounted for the growth of Yarmouth and Lowestoft as herring ports. Being on the easternmost point of England, these settlements were closest to the migration and feeding routes of the great shoals of prime pre-spawning herring in the southern North Sea. Yarmouth's autumn landings were generally larger than those of its neighbour and in the early 1900s Yarmouth gained the title of 'the herring capital of the world'. Many of the companies based in these two towns had branches in other herring areas all around the country. The result of this is that, even though this book is telling the story of a national industry, Yarmouth in particular will feature strongly.

The Herring Year

At one time the British herring industry had comprised a series of separate, localised fisheries, such as Wick in the summer and East Anglia in the autumn. At the end of these two to three-month seasons, the local fishermen would switch to catching other fish species or just haul their boats out of the water. From 1860, as fishermen started to adopt larger, safer boats, they began to try their luck at more distant herring fisheries. If an owner had invested heavily in his boat and nets it made economic sense for him to get as much use from these assets as he could. This was the beginning of the traditional migration pattern of the British herring industry.

The year began in spring around the Outer Hebrides and then the industry moved to Wick, Orkney, Shetland, Fraserburgh and Peterhead for the summer months, fishing the Buchan stock of herrings. In late summer the industry focussed on the north-east of England (from Seahouses down to Bridlington), where the herring known as the Banks group congregated before spawning. Finally the industry moved on to East Anglia for the great autumn fishery exploiting the Downs stock of herring. This fishery was the largest and also the closest geographically to the Continental markets.

There were other, smaller stocks of herring around the British coast. After Christmas, some of the fleet headed west to Plymouth, Newlyn, Milford Haven and the Irish coast – all places where herring could be found at that time of year.

The large drifter-owning companies filled any gaps in the herring year by changing their nets and fishing for mackerel. Between each season, the fishermen returned to their home ports for a week or two for boat and net repairs.

The year-round nature of the industry and the fact that herring could be found on all British coasts meant that Lowestoft and Yarmouth boats could be seen in the Western Isles of Scotland, north-east England and the south-west; Scottish boats from as far north as Shetland could be seen in East Anglia, Yorkshire, Ulster and the Isle of Man; and Cornish boats could be seen in Yorkshire and the Isle of Man. In its own way, the industry brought together all corners of the United Kingdom.

Herring Ports and Herring Stations

At the end of the nineteenth century, the large English herring stations of Scarborough, Yarmouth and Lowestoft were prosperous ports whose economy was based on a mixture of tourism, sea-trade and both pelagic and demersal fishing. However, many of the British herring fishermen came from small fishing villages in the north of Scotland, in particular from the Moray Firth area, and for these communities their entire livelihood depended on herring fishing. The growth of the Scottish herring industry had lifted many people in these remote areas out

This turn-of-the-century picture shows the Moray Firth fishing community of Cullen seeing off the herring fleet as they prepare to sail for the spring fishing around the Isle of Lewis. It could be many weeks before they would be home again. (J&C McCutcheon Collection)

Downings Bay in County Donegal was one of several remote Irish herring stations where the British herring fleet and the Scots curers worked during the early part of the year. (J&C McCutcheon Collection)

of the deep poverty caused by the Highland Clearances, some of which were still happening as late as the mid-1800s.

In Scotland, herring were originally landed and processed at small herring stations established in remote settlements; because there were few harbours, boats often had to be hauled up on to the shore. As the fishing boats became bigger and their catches became larger, it made economic sense to concentrate herring landings on the larger ports. By 1900 it was safer and more profitable for fishermen to use the larger harbours because their drifters could enter them at most states of the tide and because they had good fish markets, better facilities and access to the railways. On the distant Scottish islands, however, seasonal herring stations still sprang up each year in remote locations such as Baltasound on Shetland, Castlebay on Barra and Stronsay on Orkney.

On the Scottish mainland, the harbours at small villages that had originally been established as fishing settlements were now deserted for much of the year because the local boats worked away from home. Their harbours briefly became hives of activity in between seasons as the boats returned for maintenance and net repairs before the fleet departed for the next season. In East Anglia too, boat owners in traditional fishing villages such as Winterton and Pakefield who had once landed fish on their home beaches now based their drifters in harbour at Yarmouth and Lowestoft.

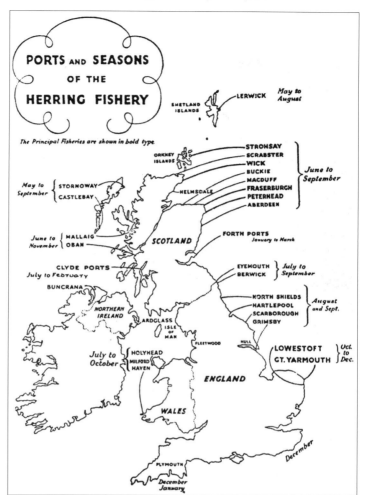

This 1930s diagram shows the different herring seasons and the main herring ports around the British coast.

Hopeman – A Scottish Herring Village

In Scotland some of the larger Moray Firth towns such as Fraserburgh, Banff, Buckie and Wick are well-known for once being herring ports but there were also a great many smaller towns and villages which also played a role in the industry. One of these is Hopeman, Morayshire, situated on the rocky southern shore of the Moray Firth, a village that was established during the 1805 Clearances. By 1855 there were forty-nine fishing boats in this village and 458 people were employed in the herring fishery as fishermen, coopers, curers, gutters and packers. In 1863, 10,000 barrels of herring were landed and packed at Hopeman, ranking it in sixth position in the whole of Scotland (although a long way behind Wick's 90,000 barrels). The village had its own salt store and an ice house. Such were the expectations of further expansion of the herring trade that in 1892 the Highland Railway built a branch

The old Ice House at Hopeman, Moray – a very impressive entrance to an underground chamber (as a small boy, the author used to imagine it was an Egyptian tomb!).

line linking the village to the main rail network and in both 1896 and 1897 over a thousand tons of fish were despatched from Hopeman by rail.

By 1900, Hopeman herring boats fished a long way from home and were landing their catches in Wick, Buckie and Fraserburgh; they travelled north in the summer to Shetland and south in the autumn to East Anglia. Hopeman was a one-industry village where very few of the men worked anywhere other than at the herring fishing and we will meet some of them as this book progresses.

The Economic Structure of the Herring Industry

In East Anglia the move to building larger vessels had resulted in a reduction in the number of family-owned boats. A new sailing drifter cost £800, which at today's retail values equates to £60,000. It required more capital than one fishing family could find from their own resources, so limited companies were formed to buy and run boats. The larger companies found that to make the most of the herring trade, it was not sufficient to just catch the fish. They needed offices at fishing ports during each season in order to support their boats, to sell their herrings and, increasingly, to manage their own in-house teams of curers and coopers. In Scotland, despite the formation of some limited companies such as the Caithness Steam Fishing Company (Wick) and the Steam Herring Fleet (Aberdeen), the majority of fishing boats remained family-owned.

At the end of the nineteenth century, anyone unconvinced about the growth of the British herring industry had only to look at recent statistics from the Ministry of Agriculture and Fisheries:

An industry that in just four years had increased production by 38 per cent and exports by 68 per cent was surely worth investing in. The value of exports in 1900 would equate today to well in excess of £200 million, giving an idea of how much the herring industry contributed to the wealth of Victorian and Edwardian Britain. This rapid expansion had, of course, been achieved at a time when the majority of the fishing fleet was still powered by sail.

There were good seasons and poor seasons. Even in a good season there were days when very few fish were landed and shore staff had no work. Sometimes the herring would be of poor quality, making it difficult to sell them in Europe. On other days

Year	TOTAL UK LANDINGS		TOTAL UK HERRING EXPORTS	
	Quantity	Value	Quantity	Value
	Hundredweight	£	Barrels	£
1897	4,993,500	1,282,000	1,119,200	1,364,000
1898	6,749,200	1,422,000	1,742,400	1,897,000
1899	5,761,500	2,006,000	1,743,000	1,899,000
1900	6,168,500	2,248,000	1,524,400	2,329,000

Table 1 UK herring landings and exports, 1897–1900.

there were huge landings which would cause quayside prices to collapse, forcing late-arriving skippers to throw away tons of perfectly good fish because no-one wanted to buy them. At these times farmers could for a few pence buy a cartload of good edible herring, which they spread on their fields as manure. So, at the end of the nineteenth century, despite the success and expansion of the herring industry, there remained an element of risk and a hint of speculation surrounding it.

As in any other industry there was tension and competition between the various sectors of the industry. Figure 2 shows the structure of the industry.

The five groups of herring buyers were all in competition to buy herrings as cheaply as they could. Fishermen were not bothered who bought their herring as long as they received a good price but they often felt that they were being diddled at the auction. Similarly, the buyers often thought that the fishermen were trying to slip some sub-standard herrings in among their catch.

Regulation of the Herring Industry

In England there was little government regulation of the trading and export of herrings – it was left to market forces, which worked well enough as long as there were plenty of buyers and plenty of herring. The government was concerned, however, about safety in the fishing industries – even as late as 1893, over 2,000 fishermen and boys had been lost at sea in one year. Various Acts of Parliament

An Overview of the Herring Trade

Boat-Owners and Fishermen

Catch and land herrings

Port Authorities, Salesmen, Auctioneers

Sell herrings on behalf of the fishermen and take
their commission

Freshers	Smoke Houses	Scotch Curers	Klondykers	Canner

These groups all compete at auction to buy
herrings. They process them and sell them on,
taking their profit.

UK Fish Wholesalers	Herring Exporters	Continental Agents

Arrange transport and sell them on to the retail trade

Fish Retailers

Display the fish and sell them to the consumer

Table 2 Diagram of the herring industry.